FLORA OF TROPICAL EAST AFRICA

ICACINACEAE

G. Ll. Lucas

Trees, shrubs or climbers. Leaves mostly alternate, sometimes opposite, coriaceous, glabrous to densely pilose, exstipulate. Inflorescence terminal, axillary or supraxillary; flowers bisexual (hermaphrodite), sometimes unisexual by abortion, regular, usually paniculate. Calyx small, 4–5-lobed, usually imbricate, rarely valvate, sometimes absent. Petals 4–5, free or united, valvate. Stamens equal in number to and alternate with petals; anthers dithecous (rarely 4-lobed), with longitudinal dehiscence; filaments often tomentose, free or adnate to petals (e.g. in *Leptaulus*). Disk rarely present. Ovary unilocular, rarely 3–5-locular (locules lost by suppression or compression), superior; ovules usually 2, pendulous, anatropous; style simple and straight or absent; stigmas (rarely 2–)3–5-lobed, often small. Fruit a drupe, unilocular, 1-seeded; endocarp woody.

A tropical and subtropical family represented by 6 genera and 12 species in East Africa, out of a world total of 58 genera and approximately 400 species.

Leaves alternate:
- Calyx and corolla present; style filiform; flowers hermaphrodite; endocarp not spiny:
  - Petals united for most of their length; filaments adnate to petals; fruits oblong to ellipsoid . . . 1. **Leptaulus**
  - Petals free or united only at base; filaments free:
    - Flowers in terminal panicles; filaments subulate; fruits with lateral fleshy appendage . . . 2. **Apodytes**
    - Flowers in supraxillary fascicles; filaments broadened towards base; fruits without lateral appendage and flattened through longitudinal axis . . . 3. **Rhaphiostylis**
    - Flowers in axillary cymes or racemes; filaments very small; fruits oblong to ellipsoid . . . 4. **Alsodeiopsis**
- Calyx absent; stigma sessile; flowers generally unisexual; endocarp spiny on inner surface . . . 5. **Pyrenacantha**

Leaves opposite . . . 6. **Iodes**

1. LEPTAULUS

Benth. in G.P. 1: 351 (1862); Sleumer in E. & P. Pf., ed. 2, 20B: 358 (1942)

Small trees or shrubs. Leaves alternate, entire, coriaceous, penninerved; secondary veins arcuate, anastomosing Inflorescence axillary or subaxillary or leaf-opposed. Flowers solitary or in subumbellate cymes, shortly pedicellate, ⚥, regular. Sepals 5, united at base, slightly imbricate. Petals 5, united almost to their apex to form a narrow tube; free portions valvate and reflexed. Stamens 5, alternating with the petals; filaments adnate to the petal for most of their length; anthers oblong-ovate. Disk absent. Ovary unilocular; style filiform, slightly eccentric; stigma small; ovules 2, pendulous. Fruit a drupe, oblong-ellipsoid; endocarp woody; seed large.

A small genus of 5 species confined to tropical Africa, two being represented in East Africa.

Flowers over 6 mm. long; petals many times longer than calyx-lobes; pedicels 2–4 mm. long; style at least 3 times as long as ovary; fruit irregularly papillose at maturity . . . . 1. *L. daphnoïdes*
Flowers small, under 4 mm. long; petals only twice as long as calyx-lobes; pedicels very short, under 1 mm. long; style equal to or up to 1·5 times as long as ovary; fruit smooth at maturity . . 2. *L. holstii*

1. **L. daphnoïdes** *Benth.* in G.P. 1: 351 (1862); F.T.A. 1: 354 (1868), pro parte; Sleumer in E. & P. Pf., ed. 2, 20B: 358, fig. 101/C–E (1942); T.T.C.L.: 251 (1949); I.T.U., ed. 2: 161 (1952); F.W.T.A., ed. 2, 1: 637 (1958); Boutique in F.C.B. 9: 261, fig. 4/A (1960). Type: Sierra Leone, Bagroo R., *Mann* 806 (K, holo.!)

Shrubs or small trees, up to 12 m. high. Young branches glabrous. Leaf-blade oblong-elliptic to oblanceolate, 6–15 cm. long, 2–6 cm. wide, acuminate with a somewhat rounded apex, cuneate, glabrous, ± coriaceous, with the midrib impressed above, prominent beneath, and with 5–7 pairs of secondary veins, conspicuous and prominent beneath, arcuate and anastomosing before leaf-margin; petiole up to 1·5 cm. long, glabrous. Flowers many, in sessile or subsessile, axillary or extraxillary cymes, fragrant; peduncle up to 3 mm. long; bracts small, lanceolate or deltoid; pedicels 2–4 mm. long, articulated near the apex. Calyx persistent, not accrescent, 2 mm. long, with deltoid lobes. Corolla white, tubular, 6–11 mm. long, with 5 small (1–1·5 mm. long) lobes at the apex. Ovary 1–2 mm. long, pubescent; style 8–12 mm. long, glabrous. Fruit ovoid to ellipsoid, ± 1·5 cm. long and 1 cm. across, beaked, irregularly papillose at maturity; endocarp stony, 10–12 mm. long, up to 8 mm. across. Fig. 1/1–3.

UGANDA. Bunyoro District: Budongo Forest, Mar. 1932, *Harris* 77 in *F.H.* 642!; Masaka District: Malabigambo, 2 Oct. 1953, *Drummond & Hemsley* 4542!; Mengo District: Entebbe, May 1906, *E. Brown* 386!
TANGANYIKA. Bukoba District: Kikuru Forest, Sept.–Oct. 1935, *Gillman* 442!
DISTR. **U**2, 4; **T**1 (around Lakes Victoria and Albert); westwards into West Africa
HAB. Lowland rain-forest and riverine forest; 1050–1225 m.

SYN. *Icacina ledermannii* Engl. in E.J. 43: 185, fig. 2/G–P (1909). Type: Congo Republic, Kasai, *Ledermann* 32 (B, holo.†)

2. **L. holstii** (*Engl.*) *Engl.* in E.J. 24: 481 (1898); Sleumer in E. & P. Pf., ed. 2, 20B: 358 (1942); T.T.C.L.: 252 (1949); Boutique in F.C.B. 9: 262, fig. 4/C (1960). Type: Tanganyika, E. Usambara Mts., Ngwelo, *Holst* 2272 (B, holo.†, K, iso.!)

Shrubs or small trees up to 5 m. high. Young branches glabrous to puberulous. Leaf-blade elliptic to oblong, 6–17 cm. long, 2·0–6·5 cm. wide, acuminate with a very characteristic linear acumen rounded at the apex, cuneate, glabrous, papery, the upper surface slightly waxy, with the midrib impressed above, prominent beneath, and with 5–7 pairs of secondary veins, prominent beneath, arcuate and anastomosing before leaf-margin; petiole 2–7 mm. long, glabrous to puberulous. Flowers many, in subsessile axillary cymes; peduncle 1–2(–5) mm. long; bracts deltoid; pedicels short, up to 0·75 mm. long. Calyx persistent, not accrescent, up to 2 mm. long, with

FIG. 1. *LEPTAULUS DAPHNOÏDES*—**1,** flowering branch, × $\frac{2}{3}$; **2,** flower, × 3; **3,** ovary, showing characteristic style, × 3. *L. HOLSTII*—**4,** small portion of flowering branch, × $\frac{2}{3}$; **5,** flower, × 3; **6,** ovary, showing characteristic style, × 3; **7,** fruit, × 2. 1–3, from *Drummond & Hemsley* 4542; 4–6, from *Verdcourt* 134; 7, from *Dawkins* 443.

± deltoid lobes. Corolla white to yellowish-green, tubular, 1·5–3·5 mm. long, with 5 very small lobes at the apex. Ovary 1–1·5 mm. long, puberulous; style up to 2 mm. long, glabrous. Fruit oblong-ovoid, 10–15 mm. long, 6–8 mm. across, turning bright red, smooth when mature; endocarp woody, 8–9 mm. long, 6–7 mm. across. Fig. 1/4–7, p. 3.

UGANDA. Mengo District: Kasa Forest, 16 Nov. 1949, *Dawkins* 443!
TANGANYIKA. Bukoba District: Minziro Forest, Nov. 1957, *Procter* 749!; Lushoto District: Bomole, 30 Dec. 1932, *Greenway* 3314!; Morogoro District: Turiani, Nov. 1953, *Semsei* 1407!
DISTR. **U**4; **T**1, 3–6, 8; Congo Republic, Angola
HAB. Lowland rain-forest; 700–1200 m.

SYN. *Alsodeiopsis holstii* Engl., P.O.A. C: 248 (1895)
A. oddonii De Wild. in Ann. Mus. Congo, Bot., sér. 5, 2: 42 (1907). Type: Congo Republic, Sanda region, *Oddon* in *Gillet* 3573 (BR, holo.!)
Leptaulus oddonii (De Wild.) Engl. in V.E. 3(2): 254 (1921)

VARIATION. Specimens from the eastern side of the Flora area tend to have longer peduncles, up to 5 mm. long.

2. APODYTES

Arn. in Hook., Journ. Bot. 3: 155 (1840); Sleumer in E. & P. Pf., ed. 2, 20B: 367 (1942)

Jobalboa Chiov., Racc. Bot. Miss. Consol. Kenya: 19 (1935)

Trees or shrubs. Leaves alternate, simple, entire, penninerved. Flowers in terminal or more rarely axillary panicles, ⚥, regular. Sepals 5, united at the base to give small deltoid lobes, persistent in fruit. Petals 5, free, valvate, oblong-linear, glabrous. Stamens 5, alternating with the petals; filaments subulate, basally attached to petal; anthers almost sagittate. Disk absent. Ovary unilocular, bearing a fleshy lateral lobe; style eccentric; stigma truncate, very small; ovules 2, pendulous. Fruit a drupe, with a large lateral appendage, and bearing remains of persistent style.

A genus of about 15 species confined to tropical and subtropical Africa, Asia and Australia (Queensland).

A. dimidiata *Arn.* in Hook., Journ. Bot. 3: 155 (1840); F.T.A. 1: 355 (1868), pro parte, excl. specim. *Schimper;* P.O.A. C: 248 (1895); Sleumer in E. & P. Pf., ed. 2, 20B: 367, fig. 103/C–E (1942); T.T.C.L.: 251 (1949); I.T.U., ed. 2: 161, t. 7 (1952); E.P.A.: 487 (1958); Boutique in F.C.B. 9: 273 (1960); K.T.S.: 238, fig. 47, t. 14 (1961); F.F.N.R.: 221 (1962); Mendes in F.Z. 2: 343, t. 72 (1963). Type: South Africa, Durban [Port Natal], *Drège* (K, syn.!)

Trees or much branched shrubs up to 25 m. high, the larger trees with fluted trunks. Bark smooth, grey; young branches glabrous to sparsely pubescent; older branches grey-brown with pale lenticels. Leaf-blade very variable, ovate-elliptic or broadly elliptic to oblong, 2–15 cm. long, 1·5–8 cm. wide, shortly acuminate or acute to obtuse, cuneate, subcoriaceous to coriaceous, the margin slightly recurved, with the midrib impressed above, prominent beneath, and the secondary veins inconspicuous, turning black on drying. Flowers many, usually in terminal panicles, rarely axillary, shortly pedicellate or sessile, sweet-scented; bracts minute or absent. Calyx small, up to 0·5 mm. long, with 5 deltoid lobes. Petals 5, free, white, drying black, linear, ± 5 mm. long and 1 mm. wide. Ovary ovoid, up to 0·7 mm. long; style eccentric, with the stigmatic surface slightly enlarged at apex, persistent. Fruit oblique, asymmetric, oblong-obovate, laterally compressed, 5–11 mm. long, 5–9 mm. high, 3–4 mm. wide, glabrous or pubescent, black with the lateral lobe red.

FIG. 2. *APODYTES DIMIDIATA* var. *ACUTIFOLIA*—**1,** flowering branch, × $\frac{2}{3}$; **2,** ovary and style, × 4; **3,** part of infructescence, × $\frac{2}{3}$; **4,** fruit, showing lateral lobe and persistent style, × 2. *A. DIMIDIATA* var. *DIMIDIATA*—**5,** part of inflorescence, × 2; **6,** flower, × 4; **7,** ovary and style, × 4. 1, 2, from *Polhill & Paulo* 1075; 3, 4, from *Faulkner* 2498; 5–7, from *Carmichael* 328.

var. **dimidiata**

Ovary pubescent; fruit sparsely pubescent. Fig. 2/5–7, p. 5.

UGANDA. W. Nile District: Koboko, Mar. 1935, *Eggeling* 1832!; Masaka District: Minziro Forest, *Wright-Hill* in *F.H.* 67!
KENYA. Uasin Gishu District: near Eldoret, Lamok R., 9 Apr. 1951, *G. R. Williams* 89!; Meru District: NE. Mt. Kenya, Marimba Forest, 14 Oct. 1960, *Verdcourt* 3001!; N. Kavirondo District: Kakamega Forest, *A. Wye* in *F.D.* 1756!
TANGANYIKA. Bukoba District: 80 km. S. of Bukoba, June 1958, *Procter* 941!; Njombe District: 14·5 km. S. of Njombe, 10 July 1956, *Milne-Redhead & Taylor* 11107!; Songea District: Matengo Hills, Mpapa, 19 Oct. 1956, *Semsei* 2539!
DISTR. **U**1, 4; **K**3–5; **T**1, 2, 4–8; Congo Republic and southwards into Angola and South Africa
HAB. Widespread in lowland and upland rain-forest and forest patches generally; 1000–2500 m.

SYN. *A. stuhlmannii* Engl. in E.J. 17: 71 (1893); T.T.C.L.: 251 (1949). Type: Tanganyika, Bukoba District, *Stuhlmann* (B, holo.†)
Jobalboa aberdarica Chiov., Racc. Bot. Miss. Consol. Kenya: 20 (1935); K.T.S.: 238 (1961). Type: Kenya, E. Aberdare Mts., *Balbo* 149 (TOM, holo.!)

var. **acutifolia** (*A. Rich.*) *Boutique* in F.C.B. 9: 274 (1960). Type: Ethiopia, Simen [Semien], Mt. Aber, *Schimper* 1315 (K, iso.!)

Ovary and fruit glabrous. Fig. 2/1–4, p. 5.

KENYA. Machakos District: Fourteen Falls, 2 Jan. 1960, *Verdcourt* 2611!; Masai District: 13 km. from Lolgorien, 15 Apr. 1961, *Glover, Gwynne & Samuel* [*Paulo*] 639!; Kilifi District: 40 km. NW. of Malindi, Marafa, 20 Nov. 1961, *Polhill & Paulo* 820!
TANGANYIKA. Musoma District: Ikizu, 2 Nov. 1953, *Tanner* 1676!; Mbulu District: SE. slopes of Mt. Hanang, 5 Feb. 1946, *Greenway* 7586!; Mbeya District: Itaka, 1 Sept. 1933, *Greenway* 3657!
ZANZIBAR. Zanzibar I., Massazini, 23 Feb. 1960, *Faulkner* 2498!
DISTR. **K**1, 3–7; **T**1–3, 5–7; **Z**; Malawi, Rwanda Republic, Ethiopia and India. It is to be expected in Zambia and probably Mozambique.
HAB. Widespread from coastal evergreen bushland to upland rain-forest; 0–2500 m.

SYN. *A. acutifolia* A. Rich., Tent. Fl. Abyss. 1: 92 (1847); Engl. in V.E. 3(2): 256 (1921)
[*A. dimidiata* sensu Oliv., F.T.A. 1: 355 (1868), pro parte, quoad specim. *Schimper*, *non* Arn. sensu stricto]
A. beddomei Mast. in Fl. Br. India 1: 588 (1875). Type: India, Mangalore, *Wight* 434 (K, holo.!)
A. bequaertii De Wild., Pl. Bequaert. 2: 79 (1923). Type: Congo Republic, Kivu, Angi, *Bequaert* 5822 (BR, syn.!, K, isosyn.!) & 5824 (BR, syn.!)
A. dimidiata Arn. subsp. *acutifolia* (A. Rich.) Cuf., E.P.A.: 487 (1958); Mendes in F.Z. 2: 345 (1963)

VARIATION (of species as a whole). Leaf-shape, often so distinctive in outline, is of no value in separating these varieties. Specimens with typically acute leaf-apices from Ethiopia, with glabrous ovaries, have their counterparts with pubescent ovaries in **T**1 and 4. There is a marked geographical trend, however, with specimens from South Africa all having pubescent ovaries (var. *dimidiata*), while as one moves northwards this material is replaced by the form with glabrous ovaries (var. *acutifolia*), until in Ethiopia only var. *acutifolia* is to be found. Where the two varieties occur side by side, intermediates may well be found in the Flora area.

3. RHAPHIOSTYLIS

Benth. in Hook., Niger Fl.: 259, t. 28 (1849); Sleumer in E. & P. Pf., ed. 2 20B: 368 (1942)

Shrubs or climbers. Leaves alternate, entire, petiolate. Flowers in ± axillary fascicles, ⚥, regular. Sepals 5, united at the base to give minute deltoid lobes, imbricate, persistent in fruit. Petals 5, free, valvate, linear-lanceolate. Stamens 5, free, alternating with the petals; filaments subulate; anthers oblong. Ovary unilocular; style eccentric, filiform, persistent;

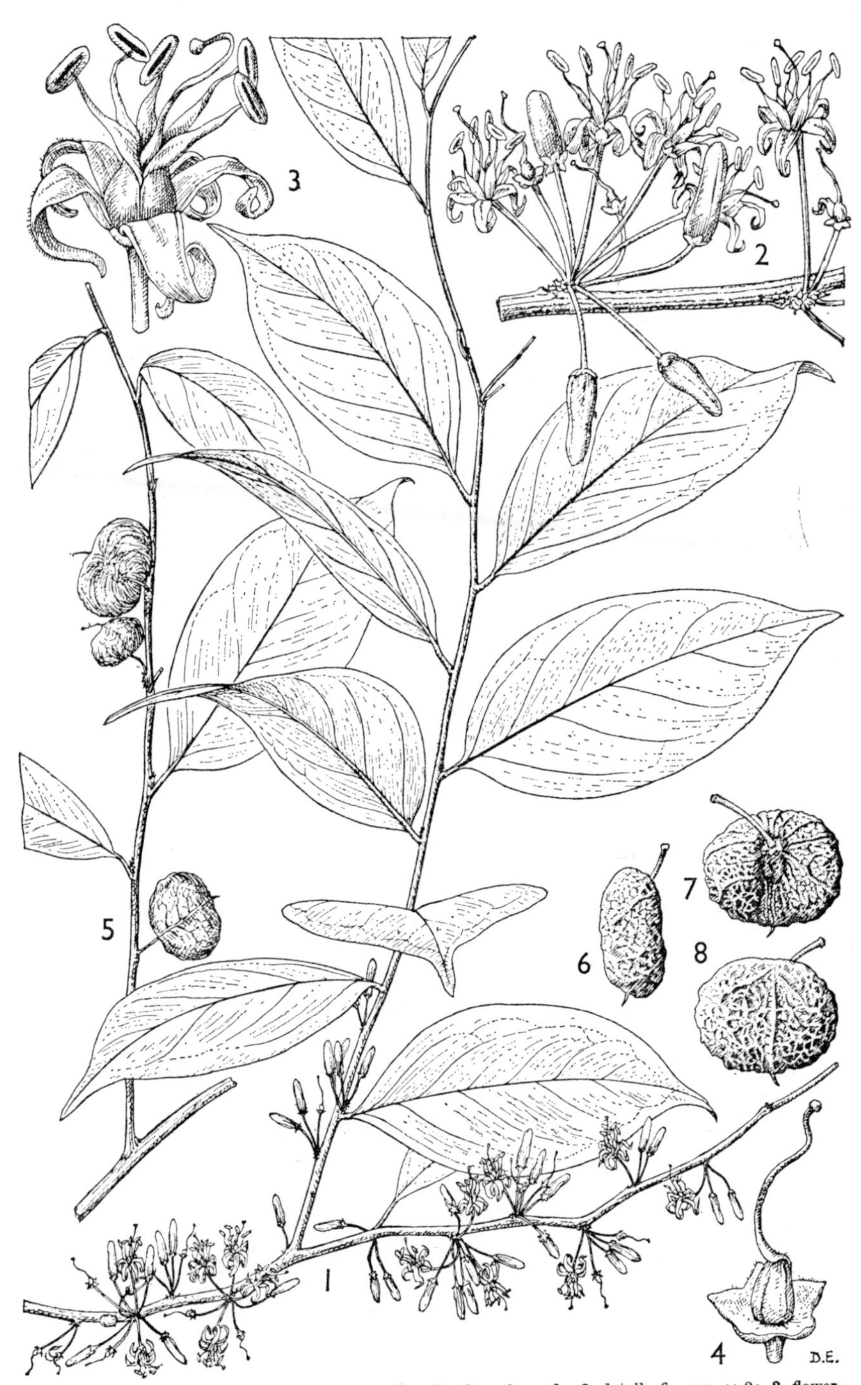

Fig. 3. *RHAPHIOSTYLIS BENINENSIS*—**1,** flowering branch, × ⅔; **2,** detail of same, × 2; **3,** flower, × 6; **4,** ovary and style, × 6; **5,** fruiting branch, × ⅔; **6–8,** fruit, viewed from various aspects, × 1. 1, from *Chandler* 1953; 2–4, from *E. Brown* 380; 5–8, from *Maitland* 881.

stigma small, capitate or discoid; ovules 2, pendulous. Fruit a drupe flattened through the longitudinal axis.

An entirely African genus of 6 species, only one of which occurs in East Africa.

R. beninensis (*Planch.*) *Benth.* in Hook., Niger Fl.: 259, t. 28 (1849); Engl. in V.E. 3(2): 256 (1921), as "*Raphiostyles*"; Sleumer in E. & P. Pf., ed. 2, 20B: 368 (1942); F.W.T.A., ed. 2, 1: 638 (1958); Boutique in F.C.B. 9: 275 (1960); F.F.N.R.: 221 (1962); Mendes in F.Z. 2: 347, t. 73 (1963). Type: Liberia, Cape Palmas, *T. Vogel* 46 (K, holo.!)

Scandent shrub or liane up to 15 m. long, often 5–6 cm. in diameter; branches glabrous, reddish-brown when young, becoming grey with age. Leaf-blade elliptic to lanceolate, 5–15 cm. long, 2–8 cm. wide, abruptly or gradually acuminate, cuneate to ± rounded at the base, glabrous, coriaceous, dark green above, paler beneath, with the midrib slightly impressed above, prominent beneath, and with 4–5(–6) pairs of secondary veins prominent on both surfaces; petiole 3–5 mm. long, canaliculate. Flowers few to many in ± axillary fascicles, pedicellate; bracts minute. Calyx 0·4–0·7 mm. long; lobes deltoid, ciliate. Petals free, white, drying black, linear, up to 7 mm. long, 1·5 mm. wide. Ovary ovoid, up to 1 mm. long, glabrous or with a few hairs at the apex; style eccentric, filiform, up to 4 mm. long, glabrous or sparsely pubescent, persistent; stigma capitate. Fruit broadly ovoid, obliquely dorsi-ventrally compressed, up to 1·3 cm. long and 2 cm. wide and 1 cm. thick, red when mature, glossy black and characteristically reticulately wrinkled on drying. Fig. 3, p. 7.

UGANDA. Toro District: Kibale Forest, Mar. 1943, *St. Clair-Thompson* in *Eggeling* 5248!; Ankole District: Mbarara, Oct. 1925, *Maitland* 881!; Mengo District: Entebbe, Oct. 1937, *Chandler* 1953!

TANGANYIKA. Bukoba District: Katoma, *Gillman* 346!; Lushoto District: Amani, 21 Dec. 1928, *Greenway* 1066!; Morogoro District: Turiani, Nov. 1954, *Semsei* 1932!

DISTR. **U**2, 4; **T**1, 3, 4, 6; westwards to Senegal and southwards to Mozambique, Rhodesia and Angola

HAB. Lowland rain-forest, riverine forest and evergreen thickets; 900–1350 m.

SYN. *Apodytes beninensis* Planch. in Hook., Ic. Pl. 8, t. 778 (1848); F.T.A. 1: 355 (1868), excl. var. β

Rhaphiostylis stuhlmannii Engl. in E.J. 17: 72 (1893), as "*Raphiostyles*"; T.T.C.L.: 252 (1949). Type: Uganda, Sese Is., *Stuhlmann* 1221 (B, holo.†, K, iso.!)

R. scandens Engl. in E.J. 43: 183 (1909), as "*Raphiostyles*". Types: Tanganyika, E. Usambara Mts., Derema, *Scheffler* 146 (B, syn.†, EA, K, isosyn.!) & *Scheffler* 165 (B, syn.†, EA, isosyn.!) & Amani, *Braun* in *Herb. Amani* 1444 (B, syn.†, EA, isosyn.!)

4. ALSODEIOPSIS

Oliv. in G.P. 1: 996 (1867) & in F.T.A. 1: 356 (1868); Sleumer in E. & P. Pf., ed. 2, 20B: 359 (1942)

Alsodeiïdium Engl., P.O.A. C: 248 (1895)

Small trees or shrubs; branchlets tomentose. Leaves alternate, entire, membranous to coriaceous. Flowers axillary in cymes or racemes, pedicellate, ⚥, regular. Sepals 5, free or partially united at the base. Petals 5, free or united up to half their length, valvate. Stamens 5, free, alternating with petals; filaments relatively short; anthers ovate-oblong or sagittate, apiculate. Ovary unilocular, hirsute; style filiform; stigma small, capitate; ovules 2, pendulous. Fruit a drupe, oblong-ellipsoid; seed large, woody.

A tropical African genus of 11 species, with one East African representative in Tanganyika.

FIG. 4. *ALSODEIOPSIS SCHUMANNII*—**1,** flowering branch, × $\frac{2}{3}$; **2,** detail of undersurface of leaf, × 2; **3,** inflorescence, × $1\frac{1}{3}$; **4,** flower, viewed from above to show fertile parts, × 6; **5,** flower, viewed from side, × 3; **6,** flower in longitudinal section, × 6; **7,** flower, showing calyx, ovary and style, other parts removed, × 8; **8,** fruiting branch, × $\frac{2}{3}$; **9, 10,** fruits, × 1. 1, 3–7, from *Faulkner* 348; 2 ,8–10, from *Drummond & Hemsley* 1755.

A. schumannii (*Engl.*) *Engl.* in E. & P. Pf. Nachtr. zu III.5: 226 (1897); T.T.C.L.: 251 (1949). Types: Tanganyika, E. Usambara Mts., Ngwelo, *Holst* 2274 (B, syn.†, K, isosyn.!) & *Holst* 2303 (B, syn.†)

Trees or shrubs up to 12 m. high. First year branches covered in golden-brown indumentum. Leaf-blade lanceolate, 5–12 cm. long, 1·5–5 cm. wide, acuminate to acute, sometimes with small mucro, rounded to cuneate at the base, membranous, ± densely strigose with golden-brown hairs, the overall appearance bicolorous, dark green above, yellow-green beneath, with (4–)6–8(–9) pairs of secondary veins prominent on both surfaces; petiole 2–8 mm. long. Flowers in lax panicles; peduncle very long (1·5–2·8 cm.), densely tomentose; bracts small, ± linear-lanceolate; pedicels 5–11 mm. long. Sepals united at the base, lanceolate, persistent, densely tomentose to villous outside. Petals yellowish, lanceolate, 4–5 mm. long, glabrous within and with a few scattered hairs mainly on midvein outside. Stamens free; filaments under 1 mm. long; anthers sagittate. Ovary subconical, densely hirsute; style filiform, up to 4 mm. long; stigma capitate. Fruit ellipsoid, slightly compressed, up to 2·2 cm. long and 1·5 cm. across, beaked at apex, ± pubescent, orange-red; pedicel elongating up to 2 cm.; endocarp woody, up to 1·5 cm. long and 1·0 cm. across, irregularly ridged longitudinally. Fig. 4, p. 9,

TANGANYIKA. Lushoto District: Amani, 26 Jan. 1931, *Greenway* 2860!; Tanga District: Muheza, 12 May 1957, *Tanner* 3500!; Morogoro District: Uluguru Mts., Bondwa Hill, 23 Mar. 1953, *Drummond & Hemsley* 1755!

DISTR. T3, 6, 7; not known elsewhere

HAB. Lowland and upland rain-forests; 900–2000 m.

SYN. *A. schumannii* Engl. in Abhandl. Preuss. Akad. Wiss.: 50, 51 (1894), *nomen subnudum*

Alsodeiïdium schumannii Engl., P.O.A. C: 248 (1895) & in E. & P. Pf. III. 5: 460 (1896)

5. PYRENACANTHA

Wight in Hook., Bot. Misc. 2: 107 (1830); Sleumer in E. & P. Pf., ed. 2, 20B: 384 (1942), *nom. conserv.*

Trematosperma Urb. in Ber. Deutsch. Bot. Ges. 1: 182 (1883)
Monocephalium S. Moore in J.B. 58: 221 (1920)

Scandent shrubs or climbers. Leaves alternate; blade entire to deeply lobed, penninerved or palmately so; hydathodes present in some species; petiole sometimes prehensile. Flowers ± axillary in spikes or racemes, dioecious, more rarely monoecious, usually unisexual by abortion. Calyx absent in ♂ and ♀. Petals of both sexes (3–)4–5(–6), united at the base, valvate, persistent. Male flowers: stamens (3–)4–5(–6), alternating with petals; filaments usually short; anthers subglobose; ovary normally replaced by a few bristle-like hairs. Female flowers: staminodes may be absent or present alternating with petals; ovary ovoid, unilocular; stigma sessile, discoid or fimbriate (variously divided); ovules 2, pendulous, 1 abortive. Fruit a drupe; exocarp fleshy; endocarp woody, bearing peg-like protuberances on the inner surface deeply penetrating the cotyledons.

A widespread genus of about 30 species, occurring throughout tropical and subtropical Africa (including Madagascar) and Asia, with 6 species in East Africa.

Stems arising from a large subglobose tuber above ground . . . . 2. *P. malvifolia*
Stems without tuber, or if tuber present then below ground:
Leaves deeply divided into 3–7(–8) lobes . . 1. *P. kaurabassana*

Leaves entire, undulate to dentate:
  Leaf-blade without marginal hydathodes . . 3. *P. staudtii*
  Leaf-blade with marginal hydathodes:
    Fruits very shortly beaked . . . . . 4. *P. sylvestris*
    Fruits with long beaks:
      Leaves sparsely hispid with appressed hairs beneath . . . . 5. *P. vogeliana*
      Leaves densely hispid beneath . . . 6. *P. sp.*

1. **P. kaurabassana** *Baill.* in Adansonia 10: 272 (1872), as "*kamassana*"; Sleumer in E. & P. Pf., ed. 2, 20B: 385, fig. 74 (1942); Mendes in F.Z. 2: 347, t. 74 (1963). Type: Mozambique, Tete, Kaurabassa, Nov. 1858, *Kirk* (K, holo.!)

Climber or twiner; stems arising from a subterranean tuberous root, dark green, drying to grey-green, hispid. Leaf-blade very variable in shape, ovate to pentagonal in outline, 4–10 cm. long, 5–15 cm. wide, subentire to more usually deeply 3–5(–7)-lobed, varying from obtuse to mucronate at apex of lobes, varying from cordate to sagittate at base, hispid above, densely hispid beneath; main veins 3, soon dividing to give the characteristic 5 or 7 palmate veins, which are continued to the margin, ending in prominent hydathodes; hydathodes ovoid, ± beaked, up to 1 mm. long; petiole 3–6(–12) cm. long, hispid. Flowers dioecious in axillary or supraxillary spikes, appearing before or during early leaf. Male flowers sessile, subtended by a small bracteole and densely compacted into an elongate spike 2–3 cm. long; peduncle 4–12 cm. long, hispid; perianth parts united at base to give 4 small deltoid lobes up to 1·5 mm. long, pubescent outside, glabrous to sparsely pubescent within; ovary replaced by a few rather coarse hairs. Female flowers fewer than ♂, sessile, subtended by a small bracteole, compacted into a more globular shape 1–1·5 cm. long; peduncle much shorter than in ♂; perianth parts as in ♂; ovary small; stigma sessile, with radiate filaments borne in an apical depression of the ovary. Fruits densely aggregated at the end of an enlarged stout peduncle up to 2·5 cm. long, ± ellipsoid, slightly compressed, 1·5–2 cm. long, 1·0–1·3 cm. thick, 5–8 mm. across, turning yellow to orange when mature, hispid.

KENYA. Northern Frontier Province: Moyale, 10 Oct. 1952, *Gillett* 14026♂! & Dandu, 16 Mar. 1952, *Gillett* 12550♀!; Machakos District: Kibwezi Plains, 23 July 1938, *Bally* in *C.M.* 7718♀ & ♂!; Mombasa District: near Nyali Bridge, 26 Jan. 1953, *Drummond & Hemsley* 1023♂ & ♀!

TANGANYIKA. Lushoto District: Korogwe, 23 Dec. 1962♂, 8 Jan. 1963 ♀, *Archbold* 103!; Tanga District: Sawa, 21 Dec. 1956, *Faulkner* 1929♀! & Kibuguni, 25 Nov. 1936, *Greenway* 4767♂!

DISTR. **K**1, 4, 7; **T**3–6; also Rhodesia, Malawi and Mozambique

HAB. Dry scrub and bushland to deciduous woodland; 10–1550 m.

SYN. *P. vitifolia* Engl., P.O.A. C: 248 (1895); T.T.C.L.: 252 (1949). Type: Tanganyika, Uzaramo District, Kidenge, *Stuhlmann* 6348 (B, holo.†)
P. menyhartii Schinz in Denkschr. Math.-Nat. Kl. Akad. Wiss. Wien 78: 427 (1905), as "*menyharthii*". Type: Mozambique, Tete, *Menyhart* 819 (K, iso.!)

2. **P. malvifolia** *Engl.* in Sitz. Preuss. Akad. Wiss. 18: 268 (1893), as "*malvaefolia*" & P.O.A. C: 248 (1895) & in V.E. 1(1): 258, fig. 255 (1910); T.T.C.L.: 252 (1949). Type: Kenya, Teita District, Ndara, *Hildebrandt* 2355 (B, holo.†); Tanganyika, Pare District, *Volkens* 2366 (BM, neo.!)

Climber or twiner up to 15 m. long; stems arising from an exposed or partially submerged swollen stem up to 1·5 m. in diameter and 75 cm. high, irregularly shaped, covered by a grey to buff often papery epidermis. Young shoots pubescent, green when young, soon becoming grey to buff and often practically glabrous, covered with a few scattered pale lenticels. Leaf-blade reniform to broadly ovate, with 3, 5, 7(–8) shallow irregular lobes, 4–12 cm. long, 4–11 cm. broad, the central lobe mucronate to emarginate depending on the position of the central vein hydathode, other lobes rounded, broadly cuneate to cordate at base, glabrous to densely pubescent; main veins 5, palmate, outer pair usually giving rise to another pair of major veins, all continued to margin and hydathode, prominent beneath, slightly raised above. Flowers dioecious, more rarely monoecious or ⚥, appearing on new shoots before or during early leaf. Male flowers sessile, subtended by a small linear bracteole, aggregated into axillary or supraxillary elongate spikes up to 6 cm. long along whole length of rhachis; peduncle non-existent; rhachis densely strigose; perianth parts united at base to give 4–5(–6) deltoid to obtuse lobes up to 1·5 mm. long, glabrous within, ± pubescent outside; ovary replaced by a few coarse hairs. Female flowers sessile, bracteolate, few on very short lateral shoots, bunched together; perianth parts 4–5, as in ♂; staminodes absent or present in the form of flattened filaments alone or less often bearing sterile anthers; ovary ovoid, densely strigose to pilose; stigma sessile, with radiate lobes. Hermaphrodite flowers having both the typical ovary and stamens are also found, often at the base of a rhachis bearing ♂ flowers at the apex. Fruits aggregated and subsessile on short lateral branches, ellipsoid, slightly compressed, 1·5–2·0 cm. long, 7–10 mm. thick, 11–14 mm. across, turning yellow on ripening, hispid.

SYN. (of species as a whole). *Trematosperma cordatum* Urb. in Ber. Deutsch. Bot. Ges. 1: 182 (1883), *non Pyrenacantha cordata* Thode. Type: plate 6 in Jahrb. d. K. Bot. Gart. 3 (1884), based on specimens grown in the Berlin Botanical Gardens from material sent by Hildebrandt about 1875 from the Somali Republic.

var. **malvifolia**

Leaf-blade not densely pubescent. Fig. 5.

KENYA. Northern Frontier Province: E. of Banessa, 23 May 1952, *Gillett* 13272⚥!; Masai District: Ol Lorgosailic, 1 Aug. 1943, *Bally* 2615♀ & ♂! & Oloibortoto, 6 Aug. 1962, *Grindlay* in *Glover* 3252A⚥!; Teita District: near Buchuma, 27 June 1961, *Verdcourt* 3190♂! & Buchuma, 16 Sept. 1961, *Polhill & Paulo* 475♀!
TANGANYIKA. Lushoto District: near Mkomazi, Mbalu Hill, 25 Jan. 1948, *Bally* 5757♀!; Tanga District: Kwale, 19 Dec. 1946, *Greenway* 7905♀!
DISTR. **K**1, 6, 7; **T**3; Ethiopia
HAB. Dry scrub and bushland; 0–1500 m.

SYN. *P. globosa* Engl. in E.J. 18: 80 (Dec. 1893), *nom. illegit.* Type: as *P. malvifolia*
P. ruspolii Engl. in E.J. 24: 483 (1898); Chiov., Res. Sc. Miss. Steph.-Paoli Som. Ital.: 51, t. 21/B (1916). Type: Ethiopia, Web Ruspoli, *Riva* 910 ♀ (FI, holo.!)

NOTE. There has been considerable confusion with this plant from N. Kenya and Ethiopia and the plant previously known as *Trematosperma cordatum* Urb. from the Somali Republic (N.). Good field-notes and more specimens now show that these are in fact the same species. I suggest that the latter may be regarded as a variety, on the difference in indumentum, but this may well prove to be artificial when a more complete distribution pattern is known.

3. **P. staudtii** (*Engl.*) *Engl.* in V.E. 3(2): 262, 264 (1921); Sleumer in E. & P. Pf., ed. 2, 20B: 385, fig. 115/A–D (1942); T.T.C.L.: 252 (1949); F.W.T.A., ed. 2, 1: 642 (1958); Boutique in F.C.B. 9: 256, fig. 2/A (1960). Type: Cameroun Republic, Johann-Albrechtshöhe, *Staudt* 568 (B, holo.†, K, iso.!)

Scandent shrub or woody climber; young stems yellow-brown, tomentose, longitudinally canaliculate; older branches practically glabrous, with a few scattered prominent lenticels. Leaf-blade oblong-elliptic to ± obovate, 7–18 cm. long, 4–9 cm. wide, apiculate to acute, rounded to subcordate at

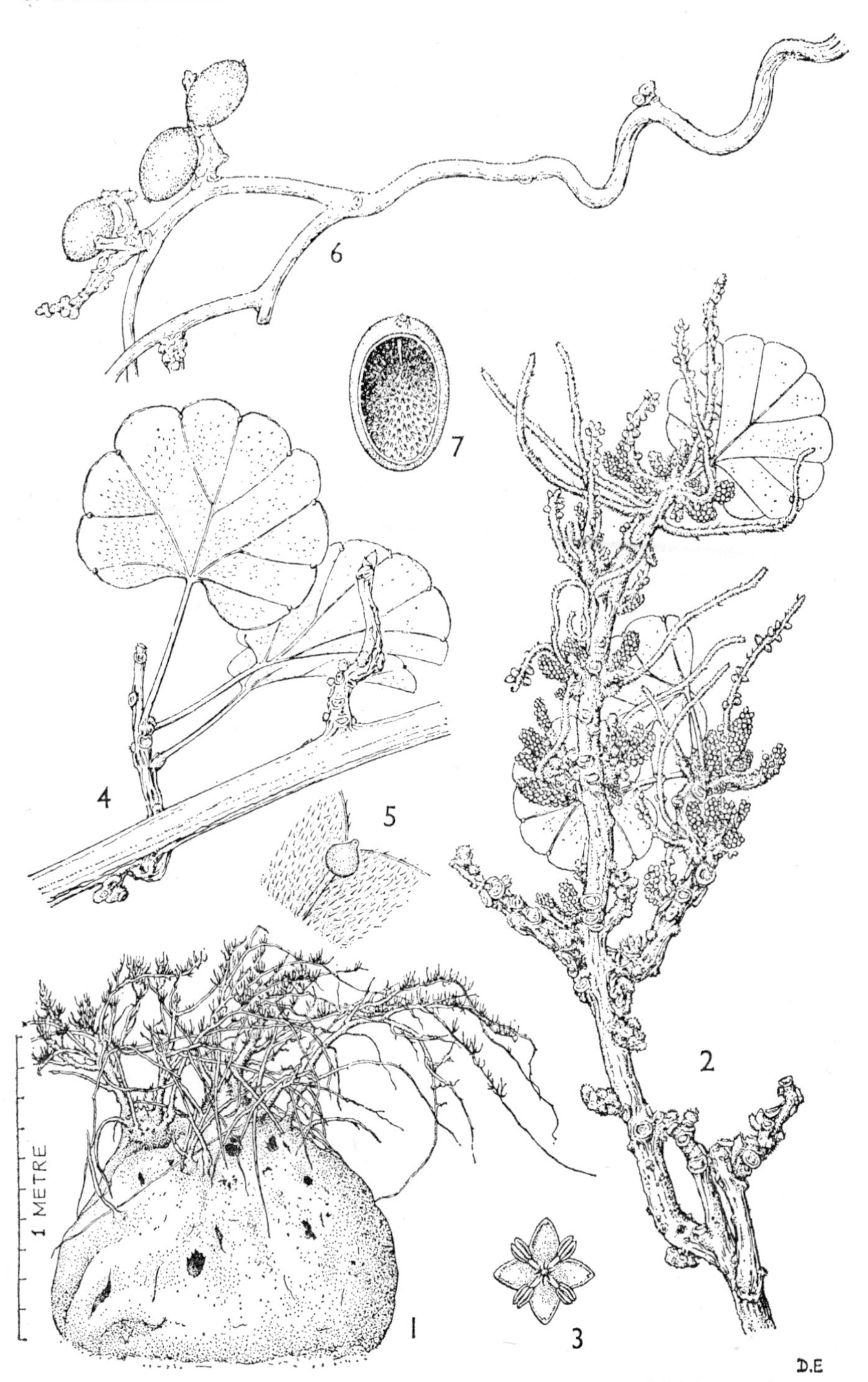

FIG. 5. *PYRENACANTHA MALVIFOLIA* var. *MALVIFOLIA*—**1**, general habit, scale shown; **2**, flowering branch, × $\frac{2}{3}$; **3**, male flower, × 6; **4**, leafy branch, × $\frac{2}{3}$; **5**, detail of hydathode, × 4; **6**, fruiting branch, × $\frac{2}{3}$; **7**, fruit cut away to show internal peg-like protuberances, × $1\frac{1}{3}$. 1, from a photograph by Bally; 2, 4, 5, from *Bally* 10570; 3, from *Verdcourt* 3717; 6, 7, from *Bally* 2615.

the base, subcoriaceous, with a repand to ± dentate margin, discolorous, practically glabrous, yellow-green to dark green, with the midrib and 5–8 pairs of secondary veins sparsely to densely grey-green to yellow-green pilose above and very prominent beneath; petiole 1·0–2·5 cm. long, deeply channelled and twisted, tomentose. Flowers dioecious, axillary or supraxillary, appearing when leaves are mature. Male flowers sessile, borne in solitary or fasciculate catkin-like spikes, densely compacted; rhachis up to 5 cm. long; peduncle almost non-existent, sometimes up to 3 mm. long, densely pilose; perianth parts united at base giving 4 oblong to linear lobes 2–4 mm. long, pubescent outside, glabrous inside; ovary rudimentary, pilose. Female flowers fewer, sessile, in solitary or fasciculate globular spikes up to 8 mm. in diameter; peduncle up to 4 mm. long, tomentose; perianth oblong-elliptic, up to 3 mm. long, enlarging to 5 mm. in fruit, persistent, pubescent outside, glabrous within; ovary tomentose; stigma radiate, filamentous, persistent. Fruit globose to ellipsoid, slightly attentuated at apex, 1–1·5 cm. long, 1–1·2 cm. across, pubescent.

UGANDA. Mengo District: Entebbe, Lake shore forest, Oct. 1931, *Eggeling* 46♀! & Entebbe, Dec. 1922, *Maitland* 583♀!
DISTR. **U4**; **T1** (*fide* Engler in V.E. 3(2): 262 (1921), no specimen seen); Congo Republic to West Africa and southern Angola
HAB. Lowland rain-forest, often at margin; 1175 m.

SYN. *Chlamydocarya staudtii* Engl. in E.J. 24: 486, t. 8/A–D (1898)
Monocephalium batesii S. Moore in J.B. 58: 221 (1920). Type: Cameroun Republic, Yaounde, Bitye, *Bates* 1277 (BM, holo.!)
M. zenkeri S. Moore in J.B. 58: 221 (1920). Type: Cameroun Republic, Bipindi, *Zenker* 4904 (BM, holo.!, K, iso.!)
Pyrenacantha batesii (S. Moore) Exell in J.B., suppl. 1: 74 (1927)
P. zenkeri (S. Moore) Exell in J.B., suppl. 1: 74 (1927)
P. ugandensis Hutch. & Robyns in K.B. 1924: 255 (1924). Type: Uganda, Entebbe, *Maitland* 459 (K, holo.!)

4. **P. sylvestris** *S. Moore* in J.B. 58: 223 (1920); Boutique in F.C.B. 9: 251, fig. 28 (1960). Type: Angola, Cabinda, Buco Zau, *Gossweiler* 6811 (BM, holo.!)

Climber up to 10 m. long; stems slender, densely pubescent, ferrugineous when young, tending to become glabrous with maturity. Leaf-blade ovate to elliptic, 6–18 cm. long, up to 10 cm. wide, acuminate, broadly cuneate to subcordate, undulate, with hydathodes present, ± discolorous, glabrous, grey-green above, densely pubescent, grey to yellowish beneath, with the midrib and ± 5 pairs of secondary veins densely hirsute and prominent beneath; petiole 1·5–4·5 cm. long, ± channelled, twisted, ferrugineous and somewhat glabrescent in older leaves. Flowers dioecious, borne in fasciculate or rarely solitary racemes, axillary or supraxillary in ♀ only; peduncle 3–10 cm. long, puberulous; pedicels 1–1·5 mm. long. Male flowers subtended by bracts up to 0·5 mm. long; perianth parts 4, united at the base, oblong to elliptic, up to 2 mm. long, glabrous within, pubescent outside; stamens with very short filaments; ovary replaced by a few short hairs. Female flowers subtended by larger bracts than ♂ (up to 2 mm. long); perianth parts as in ♂ but only up to 1·5 mm. long; ovary subglobose, densely pubescent, ferrugineous; stigma sessile, persistent. Fruit ovoid to globose, compressed, up to 2 cm. long and 1·2 cm. across, slightly apiculate, tomentose.

UGANDA. Mengo District: Mukono, Mar. 1915, *Dummer* 2439!
DISTR. **U4**; Congo Republic, Angola (Cabinda)
HAB. Lowland rain-forest over 1000 m.

NOTE. This description has been prepared mainly from central African material in conjunction with the single Uganda specimen.

5. **P. vogeliana** *Baill.* in Adansonia 10: 271 (1872); F.W.T.A., ed. 2, 1: 642 (1958); Boutique in F.C.B. 9: 254 (1960). Type: Liberia, Grand Bassa, *T. Vogel* 13 (K, syn.!)

Climber; stems slender, scabrid, often becoming glabrous at maturity, reddish-brown, longitudinally striate. Leaf-blade oblong-elliptic to elliptic, 6–22 cm. long, 3–8 cm. wide, attenuate-acuminate, truncate to subcordate or sometimes auriculate, entire to irregularly undulate, with hydathodes present, coriaceous, glabrous and glossy above, very sparsely strigose beneath; midrib slightly raised above, prominent beneath; 4–7 pairs of secondary veins slightly raised above and prominent beneath; tertiary veins similarly raised above and prominent beneath; petiole 5–20 mm. long, twisted, used for climbing. Flowers dioecious, in solitary axillary (or supraxillary in some ♀ plants) spikes; peduncle up to 5·5 cm. long in ♂ and 8·0 cm. in ♀, sparsely pubescent; bracteoles minute. Male flowers sessile; perianth-lobes 4, ± deltoid, united at the base; stamens small; ovary replaced by a few coarse hairs. Female flowers sessile; perianth-lobes as in ♂; ovary ovoid, attenuate at apex, pubescent; stigma sessile, fimbriate, persistent. Fruits clustered at the apex of an enlarged peduncle up to 10 cm. long, ovoid, attenuate at the apex to form a beak, laterally compressed, 1·5–2·2 cm. long, 1–1·5 cm. across and up to 1 cm. thick, ± pubescent, orange-red when mature.

TANGANYIKA. Pangani District: between Hale and Makinyumbe, 1 July 1953, *Drummond & Hemsley* 3119♂!; Morogoro District: 7·5 km. N. of Turiani, 31 Mar. 1953, *Drummond & Hemsley* 1940♀!

DISTR. **T3**, 6; Congo Republic and West Africa

HAB. Riverine and freshwater swamp-forests; 200–500 m.

NOTE. This very curious discontinuous distribution is no doubt partially explained by lack of material. This is true of all *Pyrenacantha* species and probably of all inconspicuously flowered climbers. However, there still remains a very large geographical gap between the Tanganyika and West African specimens.

6. **P. sp.**

Climber; stems slender, puberulous to pubescent. Leaf-blade ovate-elliptic to ± broadly elliptic, 6–13 cm. long, 2·1–6·2 cm. wide, apically attenuate, acuminate, ending in a hydathode, cordate to slightly auriculate at base, irregularly undulate, with hydathodes present, glabrous above, densely hispid beneath, with 5–7 pairs of secondary veins prominent beneath, slightly raised above. Male plant unknown. Female inflorescence a spike; flowers unknown. Rhachis in fruit up to 4·5 cm. long with some small lanceolate 0·6 mm. long bracteoles; remains of persistent perianth with ? 4 lanceolate lobes only partially united at the base. Fruits ovoid with an elongate apex (beak), laterally compressed, up to 1·7 cm. long, 1·1 cm. across, 7 mm. thick, densely pubescent, orange when ripe.

UGANDA. Ankole District: Ruizi R., 8 Feb. 1951, *T. Jarrett* 507!

DISTR. **U2**; not known elsewhere

HAB. Presumably riverine or lowland rain-forest; 1300 m.

NOTE. This is yet a further example of under-collecting. More material is urgently required to complete the picture and description of this species correctly.

6. IODES

Bl., Bijdr. Fl. Nederl. Ind.: 29 (1825); Sleumer in E. & P. Pf., ed. 2, 20B: 377 (1942)

Lianes with tendrils arising at leaf nodes. Leaves opposite, entire, penninerved. Flowers borne in axillary cymose panicles, rarely terminal umbels,

FIG. 6. *IODES USAMBARENSIS*—**1,** branch with female inflorescences, × $\frac{2}{3}$; **2,** male inflorescence, × 2; **3,** male flower, × 6; **4,** female inflorescence, × 2; **5,** female flower, × 6; **6,** detail of fruiting branch, × 2; **7,** fruit, × 6. 1, composite drawing from *Zimmermann* in *Herb. Amani* 7684; 2–7, from *Faulkner* 1045.

dioecious, pedicellate, articulated below calyx. Sepals partially united to give 4–5 deltoid lobes. Petals 3–5, ± united at the base. Male flowers with 3–5 stamens alternating with petals; filaments very short, usually flattened. Female flowers: ovary subsessile, unilocular; stigma sessile, discoid; ovules 2, pendulous. Fruit a drupe, ellipsoid; endocarp woody; seed large.

A genus of some 28 species confined to tropical Asia and Africa (including Madagascar) with but one species in East Africa.

I. usambarensis *Sleumer* in N.B.G.B. 15: 251 (1940) & in E. & P. Pf., ed. 2, 20B: 377 (1942); T.T.C.L.: 251 (1949). Type: Tanganyika, E. Usambara Mts., *Peter* 3127 (B, holo.†)

Liane; stems thin, pubescent, golden-brown when young. Tendrils usually inserted at right-angles to leaves. Leaf-blade ovate to oblong-ovate, 7·0–11·0 cm. long, 3–7 cm. wide, shortly acuminate to subacute, rounded at base, sometimes unequal-sided, paper-like, glabrous, glossy green above, glabrous to sparsely pubescent beneath with tufts of yellow-brown hairs at the junctions of the secondary veins and midrib; midrib impressed above, prominent beneath; secondary veins prominent beneath; petiole 6–10 mm. long. Male flowers: sepals (3–)4, minute, united at the base; lobes deltoid, up to 1 mm. long; petals 4, practically free to the base, glabrous within, sparsely pubescent outside; stamen-filaments short (1–2 mm. long); anthers saddle-shaped; ovary very rudimentary, usually replaced by a few hairs. Female flowers: sepals and petals as in ♂; ovary ovoid; stigma sessile, flat, discoidal, irregular in outline. Fruit subglobose, up to 1 cm. in diameter, glabrous; endocarp woody. Fig. 6.

Kenya. Lamu District: Utwani Forest, Oct.–Nov. 1956, *Rawlins* 205♂!
Tanganyika. Lushoto District: Magunga Estate, 8 Oct. 1952, *Faulkner* 1045♂ & ♀! & Longuza, 13 Dec. 1916, *Zimmermann* in *Herb. Amani* 7684♂!
Distr. **K7**; **T3**; confined to the East African coastal region
Hab. Edges of lowland rain-forest, sometimes persisting on cultivated ground; below 100 m.

INDEX TO ICACINACEAE